Table of Contents

Preface   3

The Story Behind   8

Technology. How LoRA Works   9

How the Helium Network Works   11
   How It All Started.   11
   How Helium Coin Rewarding Works.   12
      PoC Challenger   12
      PoC Challengee   13
      Witnesses   13
      Consensus Group   13
      Network Data Transfer   13
   The Math Behind HNT Distribution.   14

The Current Status and Near Future of The Helium Network   15

Helium Coin Value Growth   17

Which Helium Hotspots to Choose   18

Other Upcoming Helium Hotspot Providers   20

The Criteria to Generate the Most Revenue   20
   Hotspot Density   21
   Antenna View   24
   The Antenna DBI   24
   Minimizing Cable Loss   27

How to Buy a Hotspot using USDT   28

Buying Bobcat miner using Binance.com (or Binance.us) 28

How to Set Up Your Helium Hotspot   29
    Connecting the Helium Hotspot   30
    Placement of the Hotspot   31

How to Estimate Your Hotspot Earnings?   33
    Data transfer rewards   33
    Witness rewards   33
    Consensus rewards   33
    Challenger rewards   33
    Challenger rewards   34
    Placement   34
    Antenna view   34
    Antenna   34

How to Turn Your Helium Coins into Cash Anywhere in the World   37

How to Create an Ideal Network of Helium Hotspots   38

The Exchanges to Use to Invest in Helium Coins   39

Helium Partnerships   41

New: Helium People's Network for 5G Distribution   42

Conclusion   44

Essential Source Links   46

References   48

About the Author   51

Preface

Among all the cryptocurrency technologies, the Helium network system built to create the people's network for Internet of Things (IoT) and 5G is one of the most meaningful and robust technologies while paving the way to generating significant passive income for its miners.

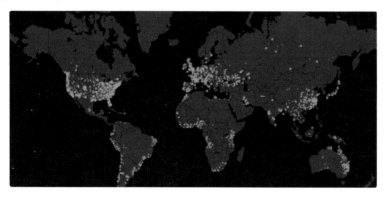
Current Helium Network coverage density map

The Helium system basically works with the Helium network created by individuals who buy and activate their Helium hotspots. The hotspots use LoRA (longfi) technology to create a network of connections among the hotspots and internet of things, and soon, the 5G network. In return, the hotspot owners are rewarded

Helium coins, which they can convert into cash using their wallets and apps.

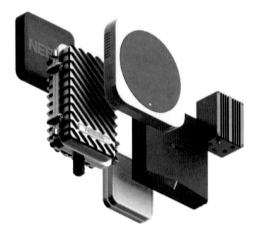

Current 3rd party Helium mining hotspots

Founded by the Silicon Valley veterans such as Shawn Fanning, Founder of Napster, and backed by Google Ventures, Unique Square Ventures and many more with an investment over $53 million, the Helium network already has a strong foundation of 42,000 Helium hotspots around the world placed by individuals like you who are generating significant passive income, as much as $10,000, from one hotspot.

Major VC companies has already supported the Helium Network

Yes, you can also become a Helium crypto miner with the best outcomes if you take the steps explained concisely in this reference book.

Helium hotspots are just like routers. They make no noise, they consume almost no energy (<$5/mo) and generate significant income for you passively.

Helium hotspot looks and feels like a router with no noise and negligible energy usage

---

Apart from being a way of generating passive income, Helium mining is also one of the best ways to enter and make an earning from the new and upcoming crypto technology - for non-technical people.

If you invest in main crypto coins or altcoins in any of the crypto exchanges, your account would be exposed to daily volatility and if you are day trading, you would need to spend so much time trading to buy low and sell high while experiencing the high stress factor.

Unboxing the Helium hotspot. It takes 10 minutes to set it up.

However, if you invest in generating income via mining Helium, then, you just spend a minimum amount to buy the Helium hotspot, about $500, for once only, and then, as you set up your Helium hotspot as explained in this book, you will gain significant income passively without even thinking about it, while possibly extending your Helium network by placing other Helium hotspots through your friends and family.

Once you read this book and apply all the key information while understanding all the key concepts, you can confidently step into Helium mining buying one or more Helium spots.

I suggest you take your time to read and digest this book. Then, you can always come back and use it as a reference.

If you are into what's interesting, new and useful, you can sign up to Innovation Party Weekly, AwayNear weekly newsletter to receive the innovations, technologies, and discounts on our radar at the moment.

Sign up for Innovation Party Weekly -
https://www.awaynear.com/innovationpartyweekly

Let's learn all about the future of The people's network of IoT & 5G in as little as 2 hours.

The Story Behind

Helium, Inc (or simply **Helium**) was founded in 2013 by Amir Haleem, Sean Carey, and Shawn Fanning to build the world's first peer-to-peer wireless network, The People's Network, to simplify connecting devices to the internet. Powered by the Helium Blockchain, Helium enables anyone to be rewarded in cryptocurrency, HNT, for becoming a network operator and providing connectivity for a new class of Internet of Things (IoT) devices.

Helium hotspot network creates the People's Network of IoT

Helium has received over $53 million since 2013 in seed, series A, series B, and series C funding. Prominent investors include: Mark Benioff, SV Angel, FirstMark, Khosla Ventures, and GV (formerly Google Ventures), Union Square Ventures, Multicoin Capital, among other VCs.

The company is based in San Francisco, California.

Technology. How LoRA Works

One of the greatest hurdles to plague any wireless industry is network coverage. That same hurdle is true for the Internet of Things (IoT). Over the last several years, Semtech has worked to create a vibrant ecosystem to drive demand for IoT applications based on its LoRa® devices and the open LoRaWAN protocol. With a permission-less, omnipresent network in the unlicensed spectrum, some of the barriers to creating and adopting a new class of low power, wide area applications have been removed.

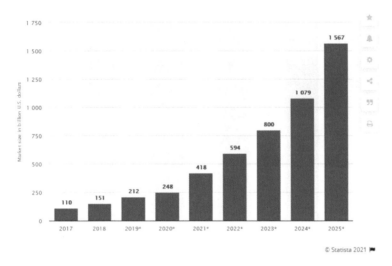

IoT projected growth 2017-2025. Source: Statista

LoRa (Long Range) is a proprietary low-power wide-area network modulation technique. It is based on spread spectrum modulation techniques derived from chirp spread spectrum (CSS) technology. It was developed by Cycleo of France and acquired by Semtech, the founding member of the LoRa Alliance and it is patented.

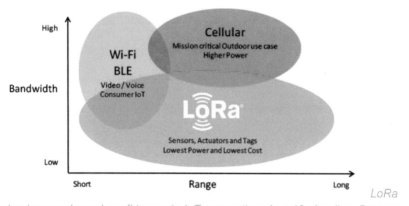

LoRa technology works as Longfi to create IoT connections in a 10mi radius. Source: Semtech

How the Helium Network Works

How It All Started.

In 2019, Helium, with its unique economic model, changed everything. By leveraging the self-built Helium blockchain, individuals are incentivized to build a new IoT network with Helium Hotspots, which simultaneously mine cryptocurrency and provide LoRaWAN network coverage for hundreds of square miles. Its unique model links these connected Hotspots together to form one homogeneous network without roaming requirements.

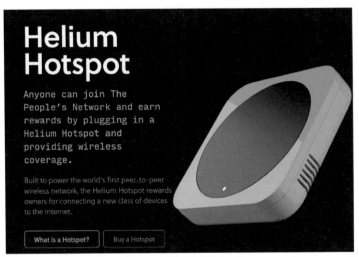

New 3rd party hotspots are released. The full list of the current ones are listed in this book. Source: Helium.com

How Helium Coin Rewarding Works.

The Helium blockchain rewards Hotspots for providing wireless coverage and verifying the Helium Network. Hotspots are rewarded in Helium Token, $HNT.

Below are how Helium rewards the hotspot miners.

PoC Challenger

PoC stands for Proof of Coverage. This is rewarded to any Hotspot that creates a valid PoC challenge and submits the corresponding receipt to the blockchain.

PoC Challengee

Awarded to any Hotspot that transmits a POC packet after being targeted by the challenger.

Witnesses

Distributed to all Hotspots that witness a beacon packet as part of a PoC Challenge.

Consensus Group

Divided equally among the 16 Hotspots that are part of outgoing Consensus Group, responsible for mining blocks.

Network Data Transfer

Distributed each epoch to Hotspots that route LongFi sensor data for sensors on the Network during that epoch.

During the course of any epoch, Hotspots are rewarded for the following list of activities:

- Submitting valid proof of coverage challenges (as a "challenger")
- Successful participation in proof of coverage as a target (as a "challengee")

- Witnessing a proof of coverage challenge
- Transferring device data over the network
- Serving as consensus group member

The Math Behind HNT Distribution.

The target rate for new HNT minted per month is 5,000,000. This means that the blockchain will produce 5,000,000 HNTs per month if it performs as designed.

The Helium blockchain consists of Epochs, which ideally has a value of 3424 HNTs/epoch.

Here is the math behind it.

Target block time is 60 secs.

Target epoch size is 30 blocks.

43,800 minutes per month / 30 minutes per epoch = 1460 epochs per month.

5M HNTs per month / 1,460 epochs per month = 3424 HNTs/epoch

Reward Type	Percentage of Reward Distribution	HNT Earned by Reward Type
PoC Challenger	.95%	32.53427
PoC Challengees	5.31%	181.849446
Witnesses	21.24%	727.397784
Consensus Group	6%	205.4796
Security Tokens	34%	1164.3844
Network Data Transfer	Up to 32.5%	Up to 1113.0145
Total	100%	3424.66

The Current Status and Near Future of The Helium Network

As of writing this book, there are over 43,500 helium hotspots in May, 2021 around the world actively mining and creating the people's network. This number was 18,000 in January 2021 and 28,000 in April 2021.

| 4,000 cities June 2021 | 200,000 hotspot backorders | 600,000 hotspots globally by 2023 |

The network effect has increased dramatically from early 2021 on

The number of hotspots and therefore, the people's network of IoT, is increasing in number worldwide dramatically because of rising awareness and temptation of generating passive income using minimal sources.

Also, the sound technology of LoRA along with the robust business model of the Helium system makes it a win-win for everyone who participates.

The Helium network coverage map. Green dots show the active Helium hotspots. Source: Helium.com

The rising number of Helium hotspots is optimally a good step for the new Helium miners since the system functions mainly using the network effect.

The Helium network has a circulating supply of 84,248,654 HNT **coins** and a max. supply of 223,000,000 HNT **coins**.

The Helium coin experiences the deflationary effect because of the halving process every two years, which in time means that the coin value will increase.

Helium Coin Value Growth

Because of the robust and meaningful technology along with the lucrative business model for both the miners and Helium Inc. itself, expectedly, Helium coin is experiencing near-exponential growth especially in 2021.

On January 1st, 2021, the value of Helium coin was $1.35 and currently, at the end of May 2021, it's $15, which is an impressive growth of 1100%.

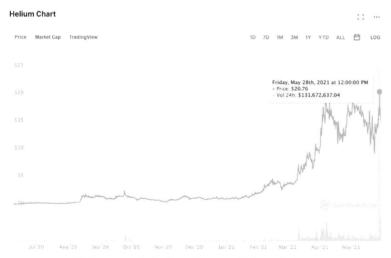

Helium growth in time. Source: Coinmarketcap

The main factors playing role in affecting the rise in value are:
- The robust LoRA technology that enables the people's network of IoT.
- Growing Helium hotspot miner population
- General public awareness and therefore, demand
- Increase in number of partners.

Which Helium Hotspots to Choose

There is mostly no distinction among the Helium hotspots with regards to their capabilities. What affects mining are outside factors such as placement, location, antenna view, etc., which will be delved into in-depth in the upcoming section.

You may want to check all the below options and choose the best one to buy, comparing the price point and delivery date while keeping in mind that all the hotspots may not be available to be shipped to your location.

Hotspots	Information
RAK Hotspot miner	Because of unprecedented demand, the number of units that can be purchased is limited to 250. You need to sign up each day and the company emails you if you are selected.

Nebra Indoor Hotspot	$350 Currently sold out. Check the shop for availability.
SyncroB.it Hotspot	$650 US available. Shipping in September
Bobcat Miner 300	$500 US available. Delivery 12-20 weeks
LongAP One Hotspot	Order soon Only available for pre-order. For UK and EU only
Kerlink	$449 Available for shopping. Ships in Fall 2021.
Freedomfi 5G	Get on the waitlist The Helium Network building the 5G-People's-Network using the Freedomfi 5G Gateways Price ranging from $1,000-5,000 depending on the signal strength/range.

Other Upcoming Helium Hotspot Providers

Name	Date	Status
HeNet BV / LongAP	2021/03/25	Approved
Smart Mimic	2021/03/25	Approved
Browan	2021/03/26	In Discussion
Dragino	2021/04/11	In Discussion
ClodPi	2021/04/11	In Discussion

The Criteria to Generate the Most Revenue

There are 4 main factors affecting the helium coin mining. They are as follows with the order of significance from the top to the bottom:
- Hotspot Density
- Antenna View
- The Antenna DBI
- Using the Right Cable

Hotspot Density

Optimum density is determined using Uber's H3 map. At the low end, hotspots won't earn from other hotspots less than 300 meters away. At the high end, hotspots can "witness" other hotspots 30 mi (50 km) out.

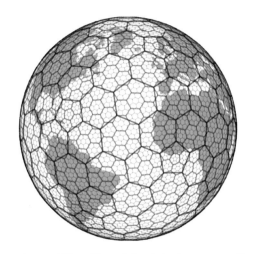

enables users to partition the globe into hexagons for more accurate analysis.

Also, the Helium network uses HIP17, which is Hex density based reward scaling.

The primary goal of HIP17 is to scale the rewards for challengees (transmitters) and witnesses (receivers) depending on the h3 hexagon they are asserted at. This change to reward distribution is to better incentivize coverage by reducing rewards earned by transmitting/witnessing hotspots in close proximity to each other.

If you want to learn all the details and get really technical, then, you can check the HIP17 documentation here.

In basic terms, if the hotspots are located optimally relative to each other, which is over 300 meter apart, the better the mining results will be.

There are 6 reward scales (1.0, 0.9, 0.8, 0.5, 0.3, 0.1) as below.

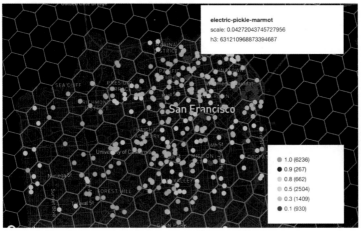

Source: Helium Engineering Blog

You may use helium.place to check any address in the world along with the active Helium hotspots, their reward scales and mining performance.

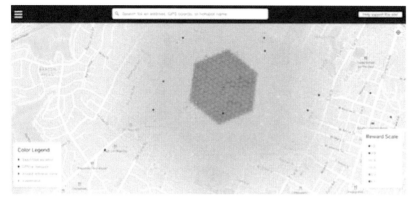

Source: Helium.place

Also, you may use the Helium Vision app and learn more technical and performance results for each location and active Helium hotspots around the particular area.

Source: Helium Vision App

Antenna View

Before you go ordering the latest and greatest super-hot antenna, make sure your antenna has a view. A "view" has three important aspects.
- Outside. Walls and even windows will block radio waves.
- High above other obstacles.
- Clear view to as much as possible.

Perfect placement. High above obstacles and close to the window. Source: Casadomo

The Antenna DBI

Many people tend to focus on the antenna DBI, which refers to the focus and shaping of the energy an antenna transmits and receives. However, a higher DBI doesn't necessarily mean higher mining outcomes.

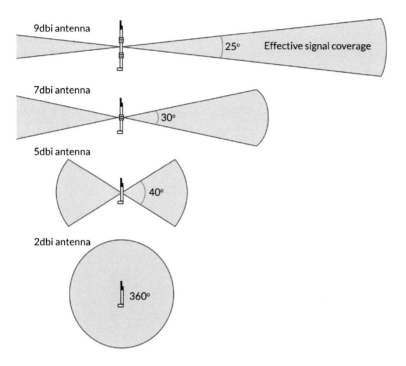

How antenna signal area changes with respect to DBI. Source: KYK13

As you can see above, a higher DBI value scans less area in the close proximity. Basically, a higher DBI means a more laser focus to a longer distance and narrower angle.

It'd therefore be considerable to have a 2 to 3 DBI antenna if you live in a populated area and if you live in a rural area and you want to focus your antenna angle to a more populated area, then, you may consider higher DBI antennas.

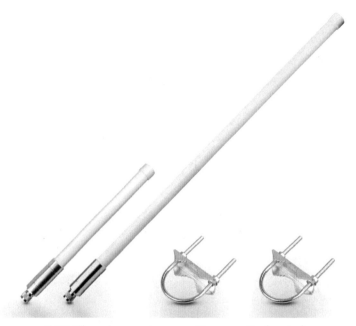

RAK Fiberglass antennas. Source: Parley Labs

Mostly, the antenna that comes with your hotspot would have 3 DBI antenna, which would be good enough for almost all cases, however, if you want to experiment with higher DBI antennas and see the Helium mining difference they make, you may want to try Parley Labs RAK antennas. There are 3, 5.8 and 8 DBI options. They are relatively affordable, there is very low risk of trying.

Minimizing Cable Loss

Normally, this is not an important factor, however, it may still have a little effect on your mining results.

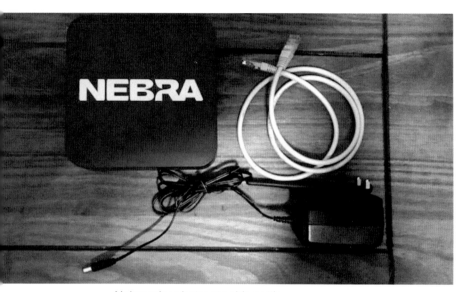

Nebra miner hotspot cable setting

It's best if you do the following with your cables:
- Minimize the cable length as much as possible, preferably less than 5 ft.
- Get a thick and clean cable.

LMR-400 coaxial cable may be a good option.

How to Buy a Hotspot using USDT

To buy a Bobcat miner, you'd need USDT, which is basically the cryptocurrency version of USD and it's a stable coin corresponding to 1 USDT = 1 USD.

You can buy USDT using the below crypto exchanges:
- Binance
- KuCoin
- Kraken
- Gate.io
- FTX
- Bitfinex
- Coinbase

You can see more information on USDT here on the Coinmarketcap website.

Buying Bobcat miner using Binance.com (or Binance.us)

For this description, we will be using the Binance.us platform to buy a Bobcat miner. Binance.com is not available in the US. Instead, Binance.us is used in the US, however, it's not available in Connecticut, Hawaii, Idaho, New York, Texas, Vermont and Louisiana. You may also use the above crypto exchanges as an alternative using the same below method to buy a Bobcat miner.

Take the below steps to buy your Bobcat miner using Binance:
- In Mugglepay, select USDC or USDT.
- Go to Binance wallet.
- Click withdraw and then, crypto.
- Click USDT.
- Under address, you can scan the barcode on the Mugglepay page.
- Ensure that TRC20 is the network selected
- Under the amount, enter the total amount including the $1 transaction fee. This additional fee is for the crypto exchange. It's important not to neglect this fee. $1 would go to Binance and the rest would go to Bobcat miner.
- Then, select withdraw and confirm.
- Enter the verification number sent to your mobile number via (Google) authenticator.
- It should take about 5 minutes for the USDT transaction to be processed.
- The process will then be complete and you will be sent a confirmation message.

How to Set Up Your Helium Hotspot

Currently, it takes about 3-4 months on average for Helium hotspots to be delivered because of high demand. However, the Helium network partners with 3rd party manufacturers as the demand keeps rising.

Once you receive your Helium hotspot, it's easy to set it up and the Helium app (Google Play | Apple App Store) will guide you all the way.

Setting up consists of two basic steps:
- Connecting the Helium hotspot
- Placement of the hotspot

Connecting the Helium Hotspot

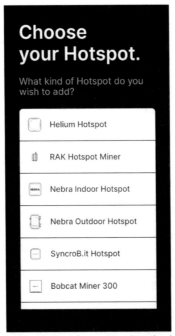

Select and connect your Helium hotspot using the Helium Hotspot app

Placement of the Hotspot

Placement 1

Open, clear view with 300+ meters from the nearest hotspot is ideal

Placement 2

Make sure there are the least number of obstacles. Basically, the window side is ideal.

Placement 3

Open, clear view with 300+ meters from the nearest hotspot is ideal

Placement 4

Any metallic disturbance isn't your friend when it comes to Helium mining.

How to Estimate Your Hotspot Earnings?

There are a number of factors affecting the results of Helium coin mining of which are listed below:

Data transfer rewards	33.31% of the 5,000,000 HNT minted every month, minus the HNT equivalent of any Data Credit purchases required to meet the 33.31% allocation, and divided between 18,174 hotspots.
Witness rewards	21.14% of the 5,000,000 HNT minted every month, divided between 34,077 hotspots, plus a bonus from redistributed HNT from data rewards.
Consensus rewards	6% of the 5,000,000 HNT minted every month, assuming your hotspot has an equal chance of being elected as the other 36,348 hotspots.
Challenger rewards	5.26% of the 5,000,000 HNT minted every month, divided between 34,077 hotspots, plus a bonus from

Challenger rewards	redistributed HNT from data rewards. 0.95% of the 5,000,000 HNT minted every month, divided between 44,982 hotspots, plus a bonus from redistributed HNT from data rewards.
Placement	Open and high place without obstacles around.
Antenna view	Number of other hotspots in the range of the antenna
Antenna	Using 3 dbi antenna is good if you live in an urban area. If you want to focus on a particular angle extending the distance towards that angle, then, you may want to use a higher dbi antenna, which is explained in the previous antenna section.

You may check **the Helium Plus earnings calculator** for your approximate earnings taking into consideration the number of hotspots in your area. However, this is only an approach.

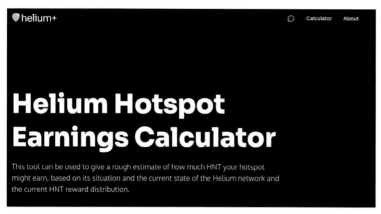

Helium.plus earnings calculator

You may check Helium.place to search for a particular address, GPS coordinate or a hotspot name. As you click on any point, you would see the name of the particular hotspot, which would let you click on the hotspot name to see the rewards of that particular hotspot.

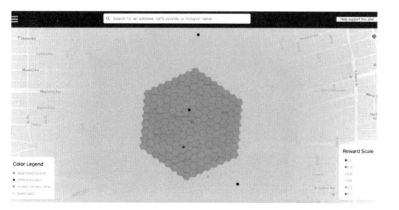

Helium.place is a great tool to give you a great idea about the possible rewards in your area.

Furthermore, you can use the Helium Vision app to see all the details of Helium hotspots in a particular area.

As you click on any hotspot, you can see the reward summary and technical information.

As you click on Recent Witnesses, you can see the witnesses with their RSSI (Recent Signal Strength Indicator)

How to Turn Your Helium Coins into Cash Anywhere in the World

For any cryptocurrency, you can go to CoinGecko and search for the crypto you are looking for to see the list of exchanges that works with that specific cryptocurrency.
As you search for Helium, you will see that there is only a limited number of exchanges.

Search for Helium in CoinGecko and scroll down to see the exchanges that process Helium

Currently, Binance.com (international), Binance.us (for the US only) and BKEX (not available for the US) works with Helium. More exchanges should include Helium as the popularity and demand increases.

How to Create an Ideal Network of Helium Hotspots

As you buy your first Helium hotspot and begin to mine, you may also want to extend your network by buying a few or a number of hotspots to place them strategically in your home location or collaborating with other people to start a Helium network in a new location.

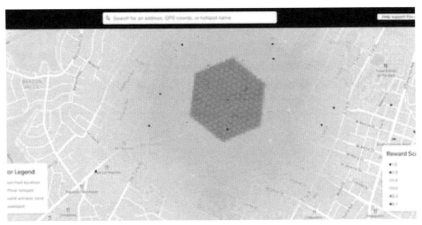

Make your research about the possible earnings from the area.

You may simply pay their internet bills or pay a portion of the earnings from the particular hotspot to start a collaboration. As you are openly saying that the hotspot works just like router spending negligible energy and it will cost nothing to them while gaining them profits, they would most likely accept your proposal.

What you need to consider:
- Make sure the place has a stable internet connection along with an open view for the antenna.
- Each hotspot needs to be at least 300 meters from each other. The range of each longfi hotspot is 25 miles (40 km).
- You may want to see the reward scale in the area using Helium.place and Helium Vision app to gain an approach regarding the earnings and reward scale in the area.

The Exchanges to Use to Invest in Helium Coins

Firstly, is it worth investing in Helium coins? Let the charts and numbers speak.

To begin with, the value of Helium coins is rising exponentially from the beginning of 2021. It was $1.3 on January 1, 2021 and it is $20 at the end of May, 2021. *That's an over 1500% increase in value.*

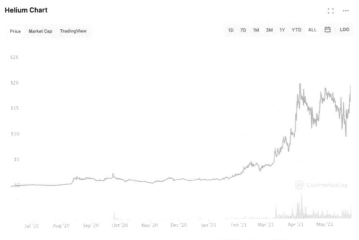
Helium value change Jul 20 - May 21

This price movement is basically determined by the network effect of users and partners.
- **User network.** Users (Helium active hotspot owners) are rewarded 5 million Helium tokens each month, which they can convert into cash. The number of users are increasing exponentially as well.
- **Partner network.** The partner network of the Helium network is extensive including Lime, Salesforce, Conserv, Smartmimic and many more. The number of partners is increasing rapidly.
- **User cases.** Helium will soon launch its Helium 5G network in collaboration with Freedomfi, which will boost its network effect and overall value as a solution provider company.
- **Meaningful.** The Helium network is the people's network connecting IoT by people, for people while incentivizing them by rewarding them with Helium tokens.

The below crypto exchanges support Helium:
- **Crypto.com** (available for the US)
- **Binance.com** (outside the US)
- **Binance.us** (for the US)
- **BKEX** (not available for the US)
- **Gate.io** (available for the US)
- **Hotbit** (available in the US)
- **WazirX** (Crypto exchange focused on the Indian market)

My personal opinion is that the Helium network will grow exponentially in the coming years, which will reflect on the price.

Helium Partnerships

There is an increasing number of company partners that use the Helium network for their IoT uses.

Below are a few of the users.

salesforce — A core value at Salesforce is trust – Helium allows them to focus on innovation for their application knowing that messages from devices are securely sent across the Helium network.

Helium-enabled devices require no cellular contracts to relay sensor data back to your

CONSERV

application. And using the standard Helium radio module, devices can be natively geolocated without the need for power-hungry cellular.

Conserv delivers real-time environmental monitoring devices running on Helium that can be used in small galleries or large art warehouses that are affordable, deliver long-range, and provide years of battery life.

New: Helium People's Network for 5G Distribution

The Helium community approved building a Helium 5G network in collaboration with Freedomfi as of April 14th, 2021.

Helium 5G Network is launching in 2021 starting from the US

This will be the first consumer-owned 5G network in the world and it will start from the US later this year. The Helium 5G network will be available internationally starting from early 2022.

The Helium 5G network will utilize The Linux Foundation Magma project as well as the open CBRS spectrum band that was made available by the FCC in January 2020.

You can see more about the technicalities of the HIP27 approval process here. You can sign up and get in the waitlist for the Freedomfi 5G gateways here.

Freedomfi 5G Gateway with the Helium Network is the First People's Network of 5G

There will be more 3rd party Helium 5G gateway manufacturers along the process as occurred with the Helium IoT hotspots.*Also, there will be modification kits to support Helium 5G Network with their existing HNT mining hotspots.*

You can read the blog post of Amir Haleem, CEO of Helium, Inc., about the Helium 5G Network here.

Conclusion

The Helium Network is a revolutionary idea that the people's hotspot network meets LoRA technology meets blockchain, which enables the global network of IoT devices and 5G.

Even though Helium, Inc. was founded in 2013, the hotspot network and Helium coin are still in its infancy and the people who take a role in building the network will not only generate passive income for themselves with a minimum investment, but they will also add high value to democratize the use of tomorrow's technologies.

Hopefully you have become as interested as we are invested into this future technology reading this book. If you have any questions or ideas, please reach us via AwayNear. As the AwayNear team, we are digital creators and we bring new ideas and projects to life. Please don't hesitate to contact us with any digital product questions as you check our website.

Essential Source Links

- **How Helium mining works** - https://www.helium.com/mine
- **Purchase a hotspot at Parley Labs** - https://parleylabs.com/
- **Helium.place** — Useful to see the value the Helium hotspot location brings - https://helium.place/
- **Helium coverage map** — This shows you where hotspots are currently - https://explorer.helium.com/
- **Original hexmap** - created to help people visualize optimal density - https://carniverous19.github.io/para1_geohip_USA.html
- **Coverage Mapping** - https://mappers.helium.com/
- **Google Earth Pro** — For looking at elevation changes around you and picking optimal spots. - https://www.google.com/earth/versions/
- **RF Line of Sight** — Helps figuring out if you have a clear line of sight to other hotspots. - https://www.scadacore.com/tools/rf-path/rf-line-of-sight/
- **Helium Vision** — Paid service w/free trial for assessing locations. - https://app.helium.vision/
- **The Hotspot** — Podcast covering the latest in Helium. - https://www.thehotspot.co/

- **Getting the right antenna -** https://store.rokland.com/pages/getting-the-right-antenna-for-your-rak-or-nebra-helium-hotspot-miner
- **Hotspots meet IoT meet Blockchain -** https://www.coinbureau.com/review/helium-hnt/
- **The Helium Network documentation -** https://docs.helium.com/use-the-network/
- **Buy Helium with a credit card -** https://www.bitdegree.org/crypto/buy-helium-hnt
- **Akash Cloud and Helium Network partnership -** https://akash.network/blog/akash-network-provides-decentralized-cloud-to-the-largest-internet-of-things-iot-network-helium
- **Top Helium Exchanges** - https://coinranking.com/coin/rGDiacWtB+helium-hnt
- **Voskcoin Youtube Channel** - https://www.youtube.com/VoskCoin
- **Tactical Investing Youtube Channel** - https://www.youtube.com/channel/UCPRC2wIfZtAlzCa_6iKE46w
- **Helium Community Signup** - https://discord.com/invite/helium

References

- **Helium network website** - https://www.helium.com/
- **How LoRA Works** - https://en.wikipedia.org/wiki/LoRa
- **The Helium Network and Semtech partnership** - https://blog.semtech.com/heliums-decentralized-approach-to-lora-based-networks
- **Helium company information** - https://www.crunchbase.com/organization/helium-systems-inc/company_financials
- **About Helium Launch** - https://techcrunch.com/2019/06/12/helium-network/
- **Helium valuation** - https://coinmarketcap.com/currencies/helium/
- **Explorer.helium website** - https://explorer.helium.com/
- **Helium Vision app** - https://app.helium.vision/
- **H3 Uber Engineering website** - https://eng.uber.com/h3/
- **Helium earnings calculator** - https://helium.plus/earnings-calculator
- **Coingecko - Crypto exchange search** - https://www.coingecko.com/
- **HIP17 Reward Scaling** - https://github.com/helium/HIP/blob/master/0017-hex-density-based-transmit-reward-scaling.md

- *Hotspot mining performance* - https://kyk13.com/a-rough-guide-to-helium-hotspot-placement/
- *RAK Hotspot Miner* - https://www.calchipconnect.com/pages/helium-compatible-miner-waitlist
- *Nebra Indoor Hotspot* - https://www.nebra.com/products/helium-indoor-hotspot-miner
- *SyncroB.it Hotspot* - https://shop.syncrob.it/products/hnt-gateway-by-syncrob-it#
- *Bobcat Hotspot Miner* - https://www.bobcatminer.com/
- *Kerlink Hotspot* - https://www.calchipconnect.com/collections/helium-compatible-kerlink-miners
- *Helium Reward Mining Concept* - https://docs.helium.com/blockchain/mining/
- *Binance* - https://www.binance.us/
- *How Antenna Gain Works* - https://www.jpole-antenna.com/2014/03/28/antenna-gain-explained
- *Antenna to Buy* - https://shop.parleylabs.com/collections/rak-accessories/products/rak-fiber-glass-lorawan-antenna-us915
- *Helium 5G page* https://www.helium.com/5G
- *Helium 5G blog post* - https://blog.helium.com/episode-two-the-path-to-5g-3f704a58661

- ***Freedomfi 5G Gateway*** - https://freedomfi.com/helium5g/

About the Author

Saygin Celen is a mechanical and industrial engineer who is passionate about design thinking and digital technologies. He has worked in automotive and renewable energy industries as an engineer. His love of design urged him to create an architectural solutions company, design and kickstart his own sneaker brand successfully and designing and consulting for machinery companies.

He is the co-founder and CEO of the digital creative agency, AwayNear and currently, he works with a top digital talent team to build digital development, business and marketing solutions for startups around the world.

He became interested in the crypto technology in early 2020 and he believes that the decentral power of the new digital currencies brings power to the people and the

Helium network technology along with the concept of the People's Network of IoT and 5G is brilliant and represent the technology - *for the people by the people.*

Made in the USA
Middletown, DE
27 September 2021